NATURAL INSPIRATIONS

space + math

The Natural Inspirations Series

Margaret Anne Evans

www.maeauthor.com

First Hardcover Edition: June 2022

First Paperback Edition: June 2022

but outer space

But outer Space,
At least this far,
For all the fuss
Of the populace
Stays more popular
Than populous .

Robert Frost

PURE MATHEMATICS IS, IN ITS WAY, THE POETRY OF LOGICAL IDEAS.

ALBERT EINSTEIN

WITHOUT MATHEMATICS, THERE'S NOTHING YOU CAN DO. EVERYTHING AROUND YOU IS MATHEMATICS. EVERYTHING AROUND YOU IS NUMBERS.

SHAKUNTALA DEVI

MATHEMATICS IS THE MOST BEAUTIFUL AND MOST POWERFUL CREATION OF THE HUMAN SPIRIT.

STEFAN BANACH

WE WILL ALWAYS HAVE STEM WITH US. SOME THINGS WILL DROP OUT OF THE PUBLIC EYE AND GO AWAY, BUT THERE WILL ALWAYS BE SCIENCE, ENGINEERING, AND TECHNOLOGY. AND THERE WILL ALWAYS, ALWAYS BE MATHEMATICS.

KATHERINE JOHNSON

MATHEMATICS IS THE MUSIC OF REASON.

JAMES JOSEPH SYLVESTER,

MATHEMATICS KNOWS NO RACES OR GEOGRAPHIC BOUNDARIES; FOR MATHEMATICS, THE CULTURAL WORLD IS ONE COUNTRY.

DAVID HILBERT

THE ESSENCE OF MATH IS NOT TO MAKE SIMPLE THINGS COMPLICATED, BUT TO MAKE COMPLICATED THINGS SIMPLE.

STAN GUDDER

THERE SHOULD BE NO SUCH THING AS BORING MATHEMATICS.

EDSGER W. DIJKSTRA

'OBVIOUS' IS THE MOST DANGEROUS WORD IN MATHEMATICS.

ERIC TEMPLE BELL

I'VE ALWAYS ENJOYED MATHEMATICS. IT IS THE MOST PRECISE AND CONCISE WAY OF EXPRESSING AN IDEA.

N.R. NARAYANA MURTHY

IT IS IMPOSSIBLE TO BE A MATHEMATICIAN
WITHOUT BEING A POET IN SOUL.

SOFIA KOVALEVSKAYA

A MATHEMATICIAN WHO IS NOT ALSO SOMETHING OF A POET WILL NEVER BE A COMPLETE MATHEMATICIAN.

KARL WEIERSTRASS

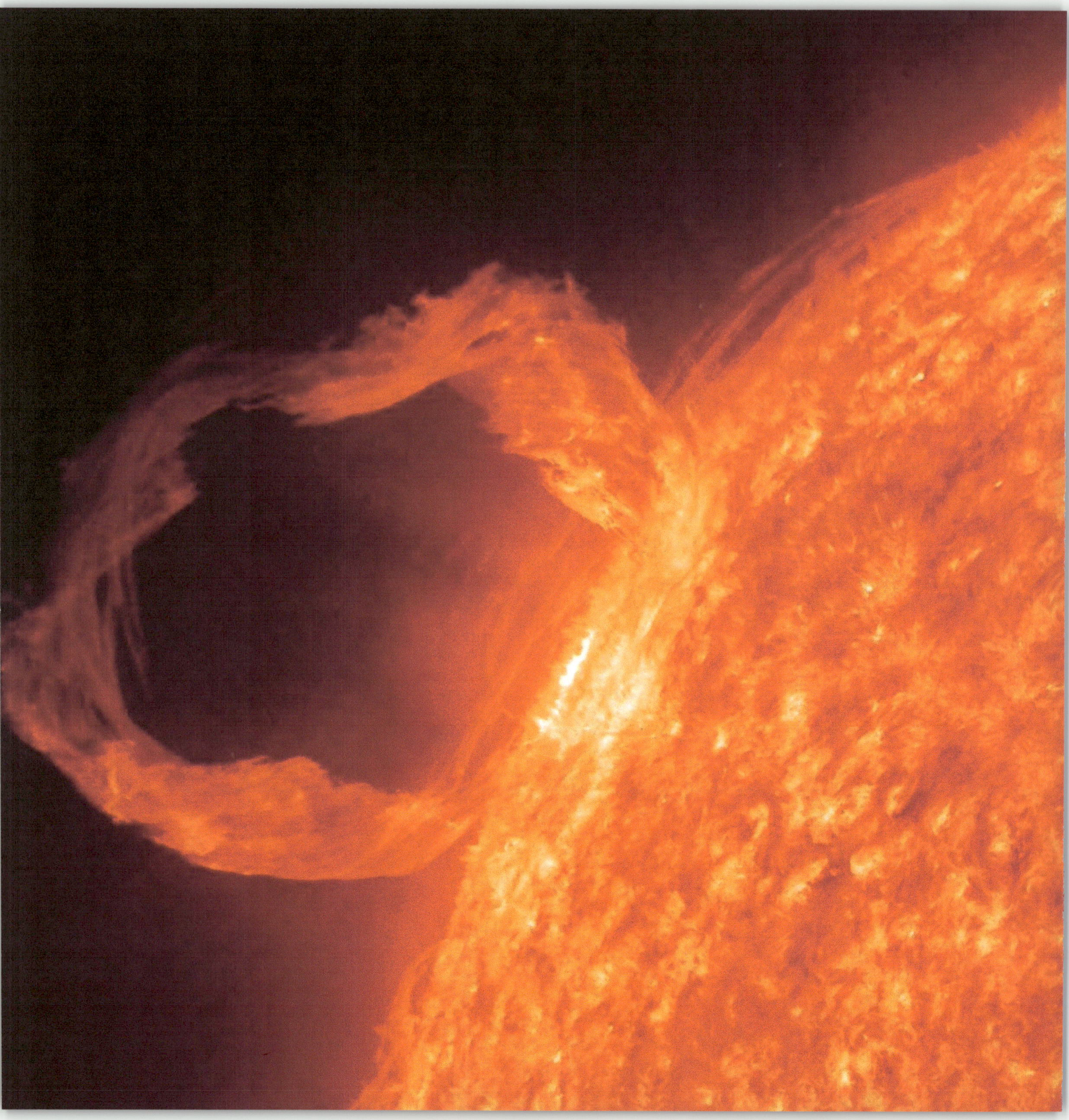

THE ONLY WAY TO LEARN MATHEMATICS IS TO DO MATHEMATICS.

PAUL R. HALMOS

MATHEMATICS COMPARES THE MOST DIVERSE PHENOMENA AND DISCOVERS THE SECRET ANALOGIES THAT UNITE THEM.

JOSEPH FOURIER

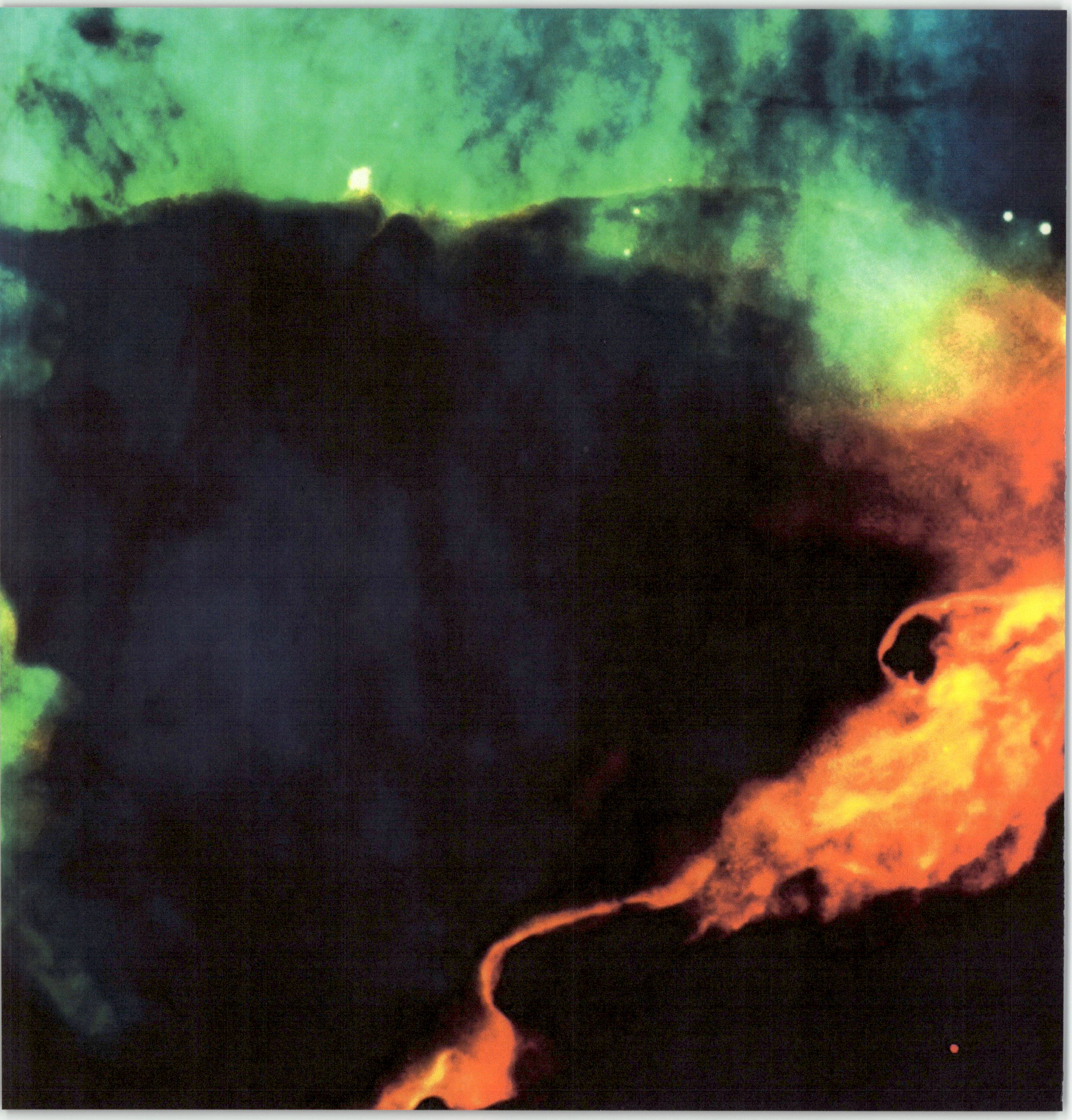

MATHEMATICS IS NOT ABOUT NUMBERS, EQUATIONS, COMPUTATIONS, OR ALGORITHMS: IT IS ABOUT UNDERSTANDING.

WILLIAM PAUL THURSTON

SOMEHOW IT'S OKAY FOR PEOPLE TO CHUCKLE ABOUT NOT BEING GOOD AT MATH. YET, IF I SAID 'I NEVER LEARNED TO READ,' THEY'D SAY I WAS AN ILLITERATE ADULT.

NEIL DEGRASSE TYSON

IN MATHEMATICS THE ART OF PROPOSING A QUESTION MUST BE HELD OF HIGHER VALUE THAN SOLVING IT.

GEORG CANTOR

IT IS CLEAR THAT THE CHIEF END OF MATHEMATICAL STUDY MUST BE TO MAKE THE STUDENTS THINK.

JOHN WESLEY YOUNG

GO DOWN DEEP ENOUGH INTO ANYTHING AND YOU WILL FIND MATHEMATICS.

DEAN SCHLICTER

I HAVE NEVER MET A MAN SO IGNORANT THAT I COULDN'T LEARN SOMETHING FROM HIM.

GALILEO GALILEI

THE DEFINITION OF A GOOD MATHEMATICAL PROBLEM IS THE MATHEMATICS IT GENERATES RATHER THAN THE PROBLEM ITSELF.

ANDREW WILES

MATHEMATICS HAS BEAUTY AND ROMANCE. IT'S NOT A BORING PLACE TO BE, THE MATHEMATICAL WORLD. IT'S AN EXTRAORDINARY PLACE; IT'S WORTH SPENDING TIME THERE.

MARCUS DU SAUTOY

TO ME, MATHEMATICS, COMPUTER SCIENCE, AND THE ARTS ARE INSANELY RELATED. THEY'RE ALL CREATIVE EXPRESSIONS.

SEBASTIAN THRUN

THE STUDY OF MATHEMATICS, LIKE THE NILE, BEGINS IN MINUTENESS BUT ENDS IN MAGNIFICENCE.

CHARLES CALEB COLTON

ONE OF THE ENDLESSLY ALLURING ASPECTS OF MATHEMATICS IS THAT ITS THORNIEST PARADOXES HAVE A WAY OF BLOOMING INTO BEAUTIFUL THEORIES.

PHILIP J. DAVIS

MATHEMATICS IS NOT A CAREFUL MARCH DOWN A WELL-CLEARED HIGHWAY, BUT A JOURNEY INTO A STRANGE WILDERNESS, WHERE THE EXPLORERS OFTEN GET LOST. RIGOR SHOULD BE A SIGNAL TO THE HISTORIAN THAT THE MAPS HAVE BEEN MADE, AND THE REAL EXPLORERS HAVE GONE ELSEWHERE.

W.S. ANGLIN

THE PURE MATHEMATICIAN, LIKE THE MUSICIAN, IS A FREE CREATOR OF HIS WORLD OF ORDERED BEAUTY.

BERTRAND RUSSELL

JUST BECAUSE WE CAN'T FIND A SOLUTION, IT DOESN'T MEAN THERE ISN'T ONE.

ANDREW WILES

MILLIONS SAW THE APPLE FALL, BUT NEWTON ASKED WHY.

BERNARD BARUCH

MATHEMATICS IS A LANGUAGE.

JOSIAH WILLARD GIBBS

THE DEFINITION OF A GOOD MATHEMATICAL PROBLEM IS THE MATHEMATICS IT GENERATES RATHER THAN THE PROBLEM ITSELF.

ANDREW WILES

I'VE ALWAYS BEEN INTERESTED IN USING MATHEMATICS TO MAKE THE WORLD WORK BETTER.

ALVIN E. ROTH

MATHEMATICS ALLOWS FOR NO HYPOCRISY
AND NO VAGUENESS.

STENDHAL

MATHEMATICS AS AN EXPRESSION OF THE HUMAN MIND REFLECTS THE ACTIVE WILL, THE CONTEMPLATIVE REASON, AND THE DESIRE FOR AESTHETIC PERFECTION. ITS BASIC ELEMENTS ARE LOGIC AND INTUITION, ANALYSIS AND CONSTRUCTION, GENERALITY AND INDIVIDUALITY.

RICHARD COURANT

THE ONLY WAY TO LEARN MATHEMATICS IS TO DO MATHEMATICS.

PAUL R. HALMOS

YOU DON'T HAVE TO BE A MATHEMATICIAN TO HAVE A FEEL FOR NUMBERS.

JOHN FORBES NASH, JR.

www.ingramcontent.com/pod-product-compliance
Lightning Source LLC
Chambersburg PA
CBHW051200220526
45473CB00003B/845
9798837887475